Tucholsky   Wagner   Zola   Scott   Sydow   Freud   Schlegel
    Turgenev   Wallace   Fonatne
        Twain   Walther von der Vogelweide   Fouqué   Friedrich II. von Preußen
                Weber   Freiligrath
                        Kant   Ernst   Frey
Fechner   Fichte   Weiße Rose   von Fallersleben   Richthofen   Frommel
        Engels   Fielding   Hölderlin
    Fehrs   Eichendorff   Tacitus   Dumas
            Faber   Flaubert
                Eliasberg   Ebner Eschenbach
    Feuerbach   Maximilian I. von Habsburg   Fock   Zweig
                    Eliot   Vergil
        Ewald
    Goethe   Elisabeth von Österreich   London
Mendelssohn   Balzac   Shakespeare   Dostojewski   Ganghofer
        Lichtenberg   Rathenau   Doyle   Gjellerup
    Trackl   Stevenson   Hambruch
Mommsen   Tolstoi   Lenz   Droste-Hülshoff
    Thoma   Hanrieder
Dach   Verne   von Arnim   Hägele   Hauff   Humboldt
    Karrillon   Reuter   Rousseau   Hagen   Hauptmann   Gautier
            Garschin   Baudelaire
        Damaschke   Defoe   Hebbel
                Descartes   Hegel   Kussmaul   Herder
Wolfram von Eschenbach   Dickens   Schopenhauer   Rilke   George
    Bronner   Darwin   Melville   Grimm Jerome   Bebel   Proust
        Campe   Horváth   Aristoteles   Federer
Bismarck   Vigny   Barlach   Voltaire   Herodot
        Gengenbach   Heine
    Storm   Casanova   Tersteegen   Gilm   Grillparzer   Georgy
        Chamberlain   Lessing   Langbein   Gryphius
Brentano   Lafontaine
    Strachwitz   Claudius   Schiller   Kralik   Iffland   Sokrates
                Bellamy   Schilling
        Katharina II. von Rußland   Gerstäcker   Raabe   Gibbon   Tschechow
Löns   Hesse   Hoffmann   Gogol   Wilde   Gleim   Vulpius
    Luther   Heym   Hofmannsthal   Klee   Hölty   Morgenstern   Goedicke
    Roth   Heyse   Klopstock   Kleist
Luxemburg   Puschkin   Homer   Möricke   Musil
        La Roche   Horaz
    Machiavelli   Kierkegaard   Kraft   Kraus
Navarra   Aurel   Musset   Moltke
    Nestroy   Marie de France   Lamprecht   Kind   Kirchhoff   Hugo
                Laotse   Ipsen   Liebknecht
    Nietzsche   Nansen   Ringelnatz
        Marx   Lassalle   Gorki   Klett   Leibniz
    von Ossietzky   May   Lawrence   Irving
                vom Stein
    Petalozzi   Platon   Knigge
        Pückler   Michelangelo   Kafka
    Sachs   Poe   Liebermann   Kock
        de Sade   Praetorius   Mistral   Zetkin
                        Korolenko

The publishing house tradition has created the series **TREDITION CLASSICS**. It contains classical literature works from over two thousand years. Most of these titles have been out of print and off the bookstore shelves for decades.

The book series is intended to preserve the cultural legacy and to promote the timeless works of classical literature. As a reader of a **TREDITION CLASSICS** book, the reader supports the mission to save many of the amazing works of world literature from oblivion.

The symbol of **TREDITION CLASSICS** is Johannes Gutenberg (1400 – 1468), the inventor of movable type printing.

With the series, tradition intends to make thousands of international literature classics available in printed format again – worldwide.

All books are available at book retailers worldwide in paperback and in hardcover. For more information please visit: www.tredition.com

tradition was established in 2006 by Sandra Latusseck and Soenke Schulz. Based in Hamburg, Germany, tradition offers publishing solutions to authors and publishing houses, combined with worldwide distribution of printed and digital book content. tradition is uniquely positioned to enable authors and publishing houses to create books on their own terms and without conventional manufacturing risks.

For more information please visit: www.tredition.com

A Description of the Bar-and-Frame-Hive With an Abstract of Wildman's Complete Guide for the Management of Bees Throughout the Year

William Augustus Munn

Imprint

This book is part of the TREDITION CLASSICS series.

Author: William Augustus Munn
Cover design: toepferschumann, Berlin (Germany)

Publisher: tredition GmbH, Hamburg (Germany)
ISBN: 978-3-8491-6549-9

www.tredition.com
www.tredition.de

Copyright:
The content of this book is sourced from the public domain.

The intention of the TREDITION CLASSICS series is to make world literature in the public domain available in printed format. Literary enthusiasts and organizations worldwide have scanned and digitally edited the original texts. tredition has subsequently formatted and redesigned the content into a modern reading layout. Therefore, we cannot guarantee the exact reproduction of the original format of a particular historic edition. Please also note that no modifications have been made to the spelling, therefore it may differ from the orthography used today.

CONTENTS

PREFACE

EXPLANATION OF THE PLATE.

THE BAR-AND-FRAME-HIVE.

HIVES AND BEE-BOXES.

THE APIARY.

THE ENEMIES TO BEES, &c.

SWARMING AND HIVING THE BEES.

ILLUSTRATIONS

PLATE I. Fig. 1.
Fig. I.
Fig. II.
Fig. III.
Fig. IV.
[pg 3]

PREFACE

Having been frequently requested to explain the use of the *bar-and-frame-hive*, in the management of bees, I have been induced to print the following pamphlet, to point out the advantages this new hive possesses over the common ones.

I have added extracts from various authorities to show the importance of transporting bees for a change of pasturage, and thus prolonging the honey harvest. Regarding the natural history of the bee, I have merely stated a few of the leading facts connected with that interesting subject, drawn from Wildman's Book on Bee-management.

London, April, 1844.

[pg 4]

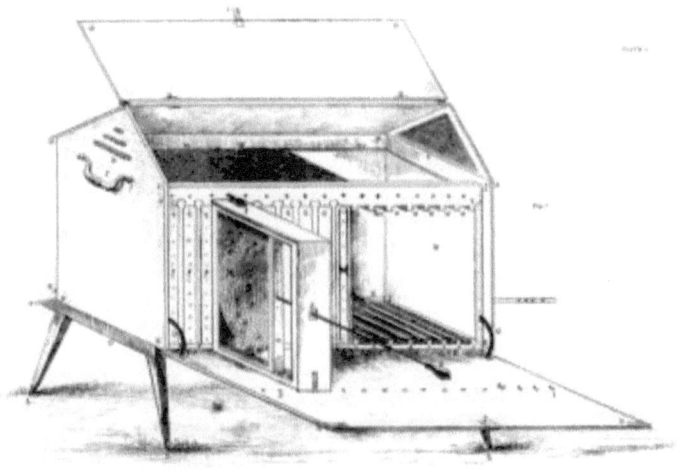

PLATE I. Fig. 1.

[pg 5]

EXPLANATION OF THE PLATE.

PLATE I, FIGURE 1.

A B C D E F and E F, the oblong box as shown in fig. 1, Plate I.

A B C D, the top lid of the oblong box; G H, the half of it made to fall back, and supported at an angle by the hinges, $h\ h$; l, the upper part of the lock of the box; $i\ k$, the two gable ends of the roof; i, the perforated zinc shown as secured in a triangular frame; and k, the outside appearance of the ventilator.

Q Q, the two quadrants, supporting the table, I J, which is formed by the side of the box, A C E E, being let down; $a\ a\ a$, &c., fifteen holes made to receive the back bolt, m, of the observation-frame, Z; $b\ b$, two bolts to fasten into the holes, c and d, when the table I J, is closed, f, being the other part of the lock.

T, one of the handles of the box (the other not seen).

U, one of the blocks (the other not shown) to keep the bottom of the box from the ground, when the four legs L L L L, are unscrewed from the four corners of the box.

X X B D, the front of the box; e, the alighting board, four inches wide, extending the whole length from F to F; X2, shows a small ledge to keep the wet from entering the bee-box, and X 1, one of the slides s, drawn out, and extending beyond the end of the box; the other half slide, s, on the *left* hand side, not drawn out in the sketch, the part under X 1, shows the opening for the ingress and egress of the bees.

[pg 6]

R, one of the two pieces of red cedar at the inside of the box, fixed at the ends, E F. E F. The Q Q, quadrants being made to work between the red cedar and the outer case or box; $v\ v$, the fillet fixed in the length of the box, on a level with the tops of red cedar; $c\ d$, the holes for the bolts $b\ b$, in the table I J.

W W, pieces of perforated zinc laid upon the tops of the bee-frames resting on the fillets, $v\ v$.

1, 2, 3, 4, 5, 6, six of the 15 grooves, half an inch deep, 9-1/2 long, and 1-1/2 of an inch broad, formed on the floor-board: the holes shown in the floor-board above the figures being made for the reception of the two pins, *a b*, in the observation-frame. No. 8, shows the "division-frame" run into the eighth groove of the floor-board, and No. 14 and 15, the bee-frames run into their respective grooves, and the 1-1/8 of an inch openings in the back closed by the slips of tin, *q q q q*, &c.

Y Y, the bar of mahogany with corresponding grooves, X X X X, &c. to those on the floor-board, at 1, 2, 3, 4, 5 and 6, and 15-2/8 holes for the top bolt, *r*, of the observation-frame, Z, to fix into. *t, t, t*, the screw nuts at the backs of the bee-frames, &c., for the screw at the end of the spindle, S, to work into, and thus hold and draw out of the grooves the bee-frames; *w*, the bee-frame containing comb and bees, drawn partly into the observation-frame, Z.

[pg 7]

A DESCRIPTION OF THE BAR-AND-FRAME-HIVE.

THE BAR-AND-FRAME-HIVE.

By first giving a general description of the "bar-and-frame-hive," the details of its construction can be better explained afterwards.

An oblong box is formed of well seasoned wood of an inch in thickness, about thirty inches long, sixteen inches high, and twelve inches broad; but the size may be varied to suit the convenience or taste of different apiarians. Instead of the lid of the box being flat, it is made in the shape of the roof of a cottage, and with projecting eaves to throw off the wet more effectually. One of the long sides of the box is constructed to open with hinges, and to hang on a level with the *bottom* of the box, and is held up by means of two quadrants. As many grooves, half of an inch broad, half an inch deep, and about 9-1/2 inches long, are formed, 1-1/8 of an inch apart, in the inside of the bottom of the box as its length will admit.

In the other side, a long half inch slip is cut for the egress and ingress of the bees, having a piece of wood about an inch thick, and four inches wide, fastened on the [pg 8] outside, just under the opening, to form the alighting board for them.

At the top, of the side of the box which is made to let down, a four inch piece of mahogany the length of the inside of the box is secured in, having corresponding grooves formed, half an inch broad, 1-1/8 of an inch deep, and half an inch apart, to those made in the bottom of the box, leaving just *twelve* inches between the bottom grooves and the upper bar grooves.

When the four legs are screwed into the four corners of the box, the small "bee-house" is ready for the reception of the "bee-frames" and the bees. The "bee-frames" are made of half inch mahogany, being twelve inches high, nine inches long, and not more than half of an inch broad, so that these frames will fit into the box, sliding into fifteen grooves formed on the bottom, and kept securely in their places by the upper grooves in the mahogany bar.

When the fifteen, or whatever number of the bee-frames intended to be used, have been run into the grooves, sheets of perforated zinc are placed on the tops of them; the 1-1/8 of an inch openings at the backs of the frames being closed with slips of tin.

One of the bee-frames is made solid, with sheets of zinc being fixed in it; this frame can then be used as a divider between any number of the bee-frames, and thus form the box into two compartments, either to augment or diminish the space in the box according to the size of the swarm, or the increasing wants of the bees for more room.

The bees are then introduced into the hive (having first closed the backs of the bee-frames with the slips of tin, and fastened the side lid of the box against them, and also removed one of the sheets of perforated zinc from the tops of the bee-frames) by dislodging the bees from the straw-hive [pg 9] in which they had been previously collected, or shaken from the boughs of the tree, where they may have settled, so as to fall upon the tops of the frames within the box; when the bees have all congregated within the bee-frames by crawling through the open spaces at the top, the perforated sheet of zinc is placed over them; the bees can then only escape through the long slip or entrance which was made for them in the front of the box.

The top lid can be closed and locked, when the bees will be secure from the gaze of the inquisitive, or the bad intentions of thieves.

Before I proceed to give any directions for the construction of the "bar-and-frame-hive" I am *anxious* to *warn* all amateur carpenters, and those who delight to superintend the labours of a "cheap working country carpenter," against the fatal error of using unseasoned wood; for, unless the "bottom board" and the "bee-frames" are made of mahogany, or some well-seasoned, hard, or close-grained wood, the advantages of the bar and frame-hive will be quite destroyed, as the great object is to have the bee-frames to slide in and out of the grooves with the *greatest facility*. Throughout the whole of the making of the hive or box, no glue should be used, unless further secured with small SCREWS OR NAILS.[1]

The oblong box, A B C D, E F and E F (Plate I, fig. 1), is to be made of well-seasoned poplar, fir, or deal, of an inch in thickness; the inside dimensions are 28 inches and 5/8 of an inch long from A to C, 10-1/2 inches broad from A to B, and sixteen inches deep from A to E.

The top lid A B C D is formed in the shape of a common roof, and made to project an inch, before, behind, and at [pg 10] the two gable ends, like the eaves of a cottage to throw off the wet.

The half of this roof G H, is made to open and fall back with hinges *h h*.

The two gable ends of the roof have holes cut in them, *i, k*, to admit the circulation of air; and secured with perforated zinc withinside to prevent the intrusion of wasps, or any other enemies to bees; the gable marked *i*, shows the perforated zinc framed into the gable, and *k* the outside appearance of the ventilator.

The side of the box marked A C E E, is made to let down and form a table I J, hung on hinges P P, and supported by the quadrants Q Q, one inch *below the level of the bottom board*.

Two handles are fixed in the ends of the box, one shown in the sketch at T.

Two blocks of wood are screwed on the bottom of the box (one shown at U) to keep it off the ground, &c., when the four legs, L L L L, at the four corners of the box are unscrewed for the convenience of packing, &c. In the opposite side or front of the box at X X, is fixed a piece of board *e*, four inches broad, and an inch thick, extending the whole length from F F; this is secured at an angle with the bottom of the box, so as to form a slightly inclined plain *e*, for the alighting board, which would be always dry for the bees to land upon. A half inch opening is made from F to F, just above the alighting board, for the ingress and egress of the bees. Slides are made *s s*, to regulate the extent of the openings, or to entirely close the entrance to the box; these slides can be drawn out when it is necessary to clean the bottom board, &c.

Within-side the box, two pieces of red cedar of half an inch in thickness, 12-1/8 inches long, 9-1/2 inches broad, are nailed on to each end at E F, and E F (one of the pieces of [pg 11] red cedar shown at R). The quadrants, Q Q, being made to work between them and the outer case. A fillet, $v\ v$, is fastened on a level with the tops of the two pieces of red cedar, to form a ledge of about a 1/4 of an inch all round, to support the sheets of perforated zinc, as shown at W W.

Sixteen pieces of mahogany, 1-1/8 of an inch broad, and half an inch deep, are to be screwed to the mahogany floor board, commencing against the piece of red cedar, R, and leaving a space between each piece, half of an inch, and finishing against the other piece of red cedar with the last; there will then be formed fifteen grooves, half of an inch in width, half an inch in depth, and 9-1/2 inches long on the floor-board as shown at 1, 2, 3, 4, 5, 6.

A bar of mahogany, Y Y, about two inches square, having grooves, X X X X, &c., corresponding to those on the floor-board, 1, 2, 3, 4, 5, 6, &c., is let in, and fastened between A and C, having a clear space of twelve inches between the floor-board, and this top bar; the object of these grooves being to receive, and keep steadily in their places, the fifteen bee-frames, when introduced into them.

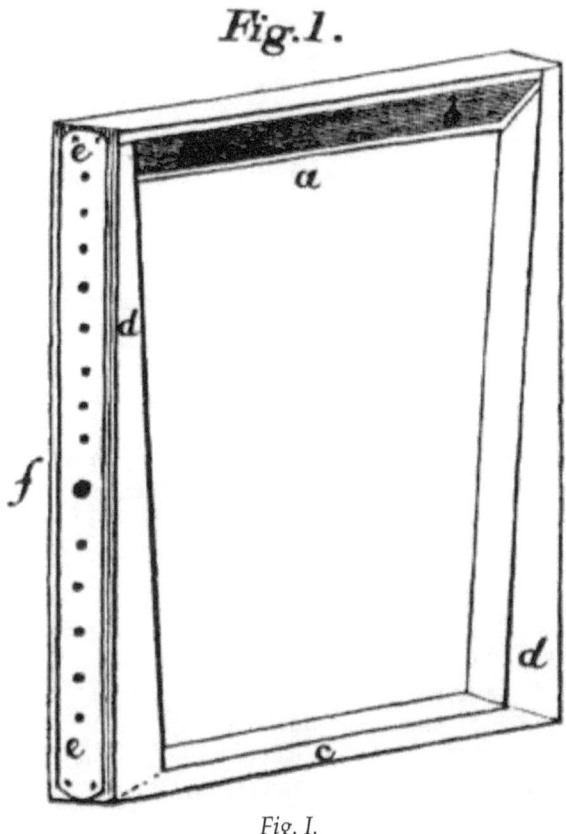

Fig. I.

The "bee-frames" are made of mahogany, nine inches long, twelve inches high, and half an inch broad. Each frame is *dove-tailed* to make it strong at the angles, and to keep it true; the upper part is formed of one inch mahogany, and *bevelled* off (as the carpenters call it) to the eighth of an inch, in the centre, as shown at *a*, fig. 1: on the two sides of this triangular bar, *b b*, pieces of glass, extending the length of the bar, are [pg 12] fixed with red lead. The two sides of the frame, *d, d*, are to increase in size, from half an inch at the top, to 1-1/2 inches at the bottom. The bottom piece, *c*, is half an inch in depth. The back of each frame has a piece of tin, about the thickness of a card, fixed on it, of the exact size, viz. twelve inches long, and half an inch broad, *e, e*. In the centre of the back of each frame, *f*, a screw-nut is let in, which is made to fit a screw at the end of a long spindle, S, Plate I, fig. 1. This spindle with a handle, Z, will screw equally well into the screw-nuts of the fifteen bee-frames and division-frame. The use of this spindle being, to draw in and out of the grooves the fifteen bee-frames when required. When the bee-frames have been put into the grooves in the box, slips of tin about thirteen inches long, and and a half broad, are slipped into their backs (being run in between the backs of the bee-frames, and the pieces of thin tin fixed upon them), to close the 1-1/8 of an inch openings. And three or four sheets of perforated zinc are laid upon the tops of the bee-frames, resting on the fillets. Thus, then, when a swarm of bees has been introduced into this box, the bees have to build their combs within the fifteen bee-frames, or whatever number may have been run into the grooves for that purpose. The bees cannot escape from above the frames, as the sheets of perforated zinc prevent them, nor from the 1-1/8 of an inch openings at the backs of the frames, as they have been closed with the slips of tin; the only open part being the long narrow slip, just above the alighting board, which was originally left for their ingress and egress.

The division-frame is made of half inch mahogany, twelve inches high, 9-1/2 long, and half of an inch broad. So that it will run into any of the grooves formed for the bee-frames; but made to fit close to the box at the end, by [pg 13] means of a slip of wood, C C, fig. 2, to prevent the bees crawling between the frame and the outer-box, as they can do round the bee-frames.

Fig. 11.

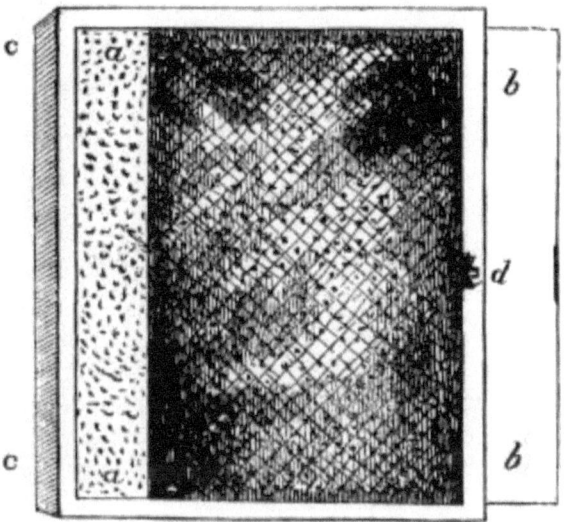

Fig. II.

The division-frame itself is closed by having two sheets of zinc run into it as shown in fig. 2, the one marked *b b b b*, and partly drawn out, being of solid sheet zinc; and *a a*, the other in the frame, of perforated zinc; *d*, being the screw-nut (like those in the bee-frames) by means of which it can be drawn out into the observation-frame, &c. Thus, wherever this division-frame is run into the bee-box, (except of course at No. 1, and No. 15 grooves) it cuts off all communication with the bee-frames on the right or left of it; and two colonies of bees may be kept in the same box, and still have distinct frames to work upon, and separate entrances, &c.

If then bees have been put into one of the bar-and-frame-hives, and sufficient time has been given them to build their combs within "the bee-frames," the frames with their contents can be drawn out into the "observation-frame," (which will be more fully described) whenever it is wished to examine the bees, &c., as the 1-1/8 of an inch spaces between the grooves will allow of a sufficient distance to be preserved, between the lateral surfaces of the perpendicular combs formed in the "bee-frames," and thus permit them to slide by each other with facility.

[pg 14]

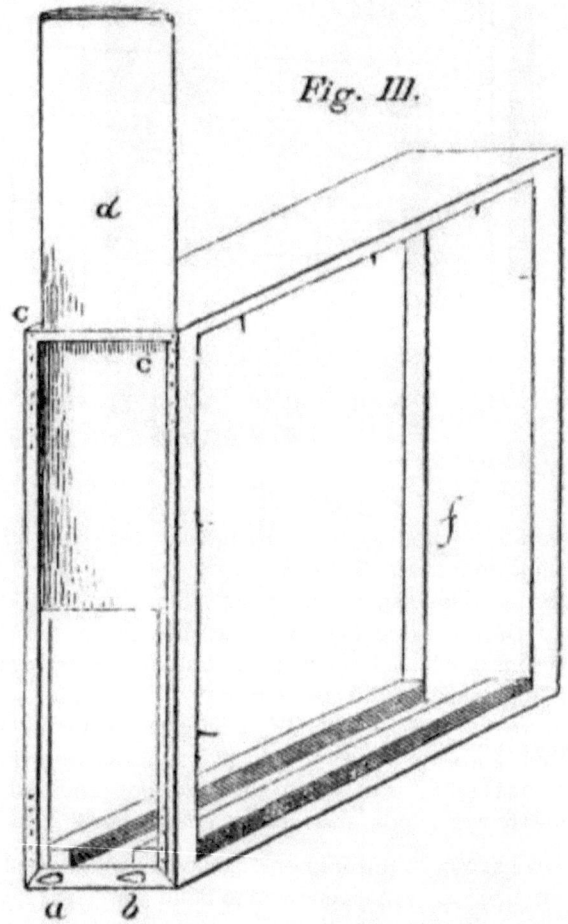

Fig. III.

The "observation-frame," fig. 3, is a mahogany frame, fourteen inches high, eleven inches long, and about four inches wide, having a single groove half an inch deep, and half an inch broad, running within its whole length of eleven inches. The two largest sides have panes of glass fixed in them with small brads. The top, bottom, and one end (this end forming the back) of this frame, are made of solid wood; the back having a small hole, *f*, 2/8 of an inch in diameter in the middle, to allow the spindle before mentioned to pass through it. The end which forms the front of the frame is open, so that any one of the bee-frames can be run into the observation-frame, but may be closed by a piece of tin (*d*) being slipt into the small grooves at *c c*. The observation-frame has two pins, *a*, *b*, to fit into the 2/8 holes made along the bottom board of the bee-box, shown by the figures, 1, 2, 3, &c., see Plate I, fig. 1, and also two small bolts *r* and *m*; *r*, the upper one to fix into the holes above X X X, &c., in the mahogany bar; (but this bolt is only used during the operation of drawing out the bee-frames into the observation frame); and the other bolt *m* at the back of the frame, to fasten into the 2/8 holes, *a*, *a*, *a*, &c., made in the lid, I J. When the two pins and the bolts of the [pg 15] observation-frame have been adjusted and fixed, the groove in it will be in a straight line with one of the grooves formed in the bottom board of the box, consequently a bee-frame can be made to slide, by means of the long spindle, in and out of the box, into the observation-frame.

The use of this "observation frame" must now be explained more fully: the top lid of the bee box, Plate I, fig. 1. G. H. being thrown up, will screen the "operator" from the bees, which are flying in and out in the front of the hive or box. The back lid, I. J., is let down, and supported by the quadrants Q. Q., and forms a table, the box having been raised from the ground by the four legs, L L L L. The observation frame is placed opposite to whichever bee-frame is to be examined; the two pins, *a*, *b*, fig. 3, running into the holes 1, 2, 3, 4, 5, &c., made in the bottom board. The small bolts, Plate I, secured at the top, as at *r*, and the back *m*: the long spindle, S, is run through the 2/8 hole in the back of the observation frame, as at Z, and the end of the spindle screwed into the screw socket *t*, at the back of the bee-frame *w*; the two pieces of tin on the right and left of the bee-frame are pulled out (of course the observation frame being empty, and having the piece of tin from its front taken out), the operator holding by the handle, *z*, of the spindle, gradually draws out the bee-frame into the observation frame, and after examining the bees and comb, gently returns the bee-frame into its groove in the floor-board: the two slips of tin are then replaced in the backs of the bee-frames: the spindle is unscrewed and withdrawn, the bolts are unfastened, the observation frame being kept firmly in its place, held by the left hand of the operator, whilst with the right he runs in the long slip of tin, d, fig. 3, into the front of the observation frame, to keep the bees (escaped from the returned bee-frame), until the [pg 16] observation frame is again fixed opposite to another bee-frame, when the tin is withdrawn and the bolts fastened as before. It has been shown that by these means, each bee-frame, and the bees and comb contained in it, can be easily drawn out and examined, without interfering with any other part of the hive, or occasioning the loss of a single bee.

The whole of the interior of the hive is thus open to inspection at any moment, and a choice can be made of the combs containing the most honey, or the bee owner enabled to trace the devastation of the honey moth, and ascertain the presence of any other enemy, and this without the assistance of smoke, which must be injurious both to the bees and their brood.

When the bee-frame is returned and secured, the observation-frame is removed; then the lid, I J, being shut up and bolted, and the upper lid, G H, closed, the box may be locked up. When the bees have been shut in with the slide in the front, the hive or box is ready to be transported anywhere, to procure new pasturage for them, which, as every experienced bee-keeper knows, is of the greatest benefit to prolong their honey-harvest.

Perfect protection from wet and the vicissitudes of temperature, is partly ensured by the external bee-box being made of well-seasoned wood; poplar is recommended as of a looser grain than fir, deal, &c., and consequently, not so great a conductor of heat; but the objection to wooden bee-hives or boxes, for being more easily affected by the variations of the temperature, is removed by the construction of the "bar frame-hive;" for the bee-frames form, as it were, a smaller box within the oblong box, and are not in immediate contact with the external air, but have a half inch space nearly all round them, which will to a certain [pg 17] extent maintain an equable temperature for the bees, both in summer and winter.

Any moisture condensed from the heated air generated by the bees, is carried off through the perforated sheets of zinc above the frames, and cool store-room for the honey is also thus secured.

A feeding trough is made on the principle of a bird-glass: with a tin feeder and a small bottle for the liquid food to be put into.

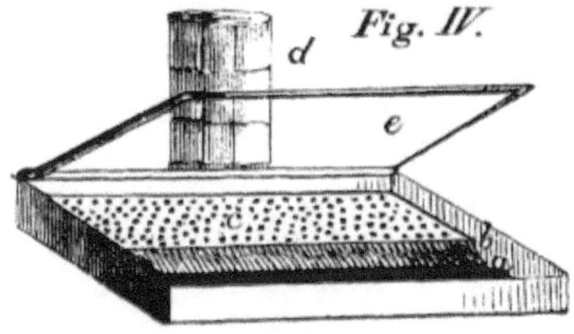

Fig. IV.

The tin feeder is six inches by 7-1/2 long, and one inch deep, and just fits on to the top of the bee-frames, where the perforated sheets of zinc are laid; within this feeder a half inch opening is cut at the bottom, fig 4, *a*, and an inclined plane *b*, reaching half way up the depth of the trough; and a sheet of perforated tin, *c* (placed horizontally from point *b*,) through which the bees suck the food, which is kept at the same level by atmospheric pressure; for as the food is drawn down below the mouth of the bottle, *d*, air forces itself into the bottle, and the same quantity of food trickles down into the feeder, a piece of glass, *e*, exactly the same size as the feeder, is placed over it, through which the bees may be seen whilst feeding, and the feeding trough will be nearly of the same temperature as the interior of the box or hive, and prevent the bees being chilled, as they would be in winter, if compelled to descend for their food; and besides, the bees are less likely to be attacked by wasps or strange bees when fed from above, as the intruders would have to ascend through the mass of bees in the box, which would be attended with danger to them.

[pg 18]

The bees can be fed when necessary by one of the sheets of perforated zinc being drawn on one side, and the feeding trough, with the bottle of food in it, being placed over the opening; when the bees will ascend through the half inch space at *a*, and feed themselves with the liquid, or carry it away and store it up for future use.

[pg 19]

HIVES AND BEE-BOXES.

Having given a description of the bar-frame-hive, it will be as well to enter into the comparative advantages of using wooden boxes and straw hives.

Some apiarians confine themselves to the use of straw hives, others to wooden boxes, and a third party use both; but as far as the bees are concerned it matters little what kind of hive is given them, for if the season be favourable, and the bee-pasturage rich with flowers, they collect and store up the honey in their combs in any receptacle of any shape or size, provided it affords them shelter from the weather.

Hives made of straw are generally preferred for an out-of-door apiary, as being less liable to be over-heated by the rays of the sun, and in the winter they exclude the cold better than hives made of other materials, while the moisture arising from the bees is more quickly absorbed within the hive, and does not run down the sides as it generally does in wooden hives or boxes; at the same time they are always to be obtained from their cheapness, and from their simplicity easily understood and made use of; wooden boxes can only be used with advantage in a bee-house, they stand firmer on the bottom boards, or one upon another, they admit of having glass windows, through which to observe the operations of the bees, and they are not so liable to harbour moths, spiders, and other insects, as the straw hives.

[pg 20]

The objects to be attained in the construction and management of an apiary, are, to secure the prosperity and multiplication of the colonies of bees, to increase the amount of their productive labour, and to obtain their products with facility, and with the least possible detriment to the stock. It is to the interest of the owner, therefore, that he provide for the bees shelter against moisture, and the extremes of heat and cold—especially, sudden vicissitudes of temperature, protection from their numerous enemies, every facility for constructing their combs and for rearing their brood, and that the hive should be so constructed as to allow of every part of the combs to be inspected at any moment, and capable of removal when requisite: and while attention is paid to economy, it should be made of materials that will secure its durability.

These observations apply equally to the straw hives, boxes, or whatever the bees may be lodged in or hived. Some cultivators of bees have been chiefly anxious to promote their multiplication, and to prevent the escape of the swarms in their natural way, by forming artificial swarms, by separating a populous hive previous to its swarming, into two parts, and allowing to each greater room for the construction of their works. Others, and the most numerous class, have contemplated only the abundance of the products which they yield, and the facility of extracting them from the hive, without showing any particular solicitude as to the preservation of the bees themselves. Another class of apiarians have, on the other hand, had it more particularly in view, to facilitate the prosecution of researches in the natural history and economy of bees.

Then, again, amongst apiarians a diversity of opinion exists regarding the system to be adopted in the management of the hives, whether the bees are to be kept in single [pg 21] hives, caps or bell-glasses, and extra boxes, which may be added at the top, which is called the *storifying* system; or inserting additional room at the bottom, called *nadering*; or whether adding boxes at the sides, called the *collateral* system, should be followed out; and a plan of ventilating the boxes has been added to the last system, but experience has proved that it is utterly useless, as in spite of ventilating tubes and thermometers, the bees have swarmed, and the queen-bee has deposited her eggs in the collateral boxes and destroyed the purity of the honey.

No successful plan has been yet devised to ventilate the combs where the bees cluster; for the bees prevent the circulation of the cold air amongst the combs by immediately forming themselves in thick rows at the bottom of the combs; and instead of ranging the fields to gather honey or pollen, have to collect together and idle away their time to retain the necessary heat for the formation of the combs, or to rear their brood.

As a single hive, Huber's leaf-hive is certainly the best; but it requires great attention, and none but experienced apiarists can use it for the purpose of trying experiments; but in the hands of experienced apiarists it is invaluable. All other single hives are objectionable, as neither the proceedings of the bees can be observed, nor the honey taken out, without either destroying the bees, or driving them out with smoke by which much of the brood is killed; or if rainy weather occur at the time the bees are preparing to throw off a swarm, and the hive be filled to its utmost limits with comb, all the bees must remain idle till the return of fine weather for want of room.

To meet this objection, some apiarians have straw-hives with flat wooden tops made, or use boxes, and have holes cut in them at the top, so that small glasses may be added, [pg 22] when the bees require room. But this does not prevent swarming, and besides, the flatness of the roof is prejudicial, as it allows the moisture which exhales from the bees to collect in the roof, and to fall in drops at different parts, to the great injury of the subjacent contents of the hive, and, like the common straw hive or square box, the bees cannot be examined, except partially through the windows made in the sides.

To remedy this evil, the further plan of *storifying* hives or boxes, was introduced, and by this method swarming may to an extent be prevented, and the wax and honey can be taken without destroying the bees; and with the same view was introduced the *collateral* system, which is adding room at the sides (of course preserving a free communication between the boxes and hives). But there are objections to the *collateral* system, as it is now a very well established fact, that partitions of any kind are detrimental to the prosperity of the bees; and the same applies, though perhaps in an inferior degree, to the *storied* system, or hives and boxes divided into stories one above another; besides that which holds good equally to all hives or boxes, that it is not possible to proportion the hives in all cases to the magnitude of the swarms, or the energy with which they labour.

In single hives the honey becomes bad and discoloured from being put into the old breeding cells. In double storied, or collateral hives, the bees are divided, and live in different families; while their own preservation, and that of the brood, requires them to live in the strictest union; the heat also necessary for the secretion of wax is lessened by the division of the bees into different groups. And, besides, all these different hives or boxes should have some [pg 23] sort of protection from the weather, either in the way of eaves or covers, or be placed in a shed or bee-house.

They require also centre boards and division tins, &c. to separate one hive or box from another, floor boards for them to stand upon, as well as stands or stools to raise them from the ground, &c., for a description of which, and a full history of all hives and boxes, I refer the reader to Dr. Bevan's "Honey-bee."

In mentioning the defects of these different boxes and hives, I do not mean to condemn them as useless, for they will all answer to a certain extent the purposes for which they were intended, rewarding the attentive bee-keeper, according to the seasons, and enabling the bees to send forth many swarms, and collecting and storing up their treasures of honey; but my object has been to point out briefly to those anxious for the better, more extended, and economical mode of bee-management, the difficulties to be provided against, and to recommend to their consideration the advantages offered in the bar frame-hive. But, however, I should not be doing justice to Mr. R. Golding, if I did not particularly mention his "improved Grecian hive" by the use of which combs may be removed from the interior of the hive and inspected at pleasure: this improvement he has effected by carefully investigating the laws of the insects for whose use the hives were intended, and by a particular arrangement of the bars, (every alternate one being furnished with guide combs,) the bees have been induced, in a manner at once simple and beautiful, to construct a uniform range of combs. When the hive is filled with honey, two or three, or more of the bars may be, at any time, removed, or exchanged for unoccupied bars, without much disturbing the brood combs, all annoyance from the bees being prevented [pg 24] by a whiff or two of tobacco smoke being blown into the hive at the time of the removal of the bars. With the protection of a bee-house these hives can be applied to many of the systems of bee-management, and prove equally profitable, and more manageable than some of the newly-invented hives.

[pg 25]

THE APIARY.

Next of importance to the kind of hive and the system to be followed, is the proper situation of an apiary. This subject engaged the attention of bee-keepers in ancient as much as in modern times; but the directions given by Columella and Virgil are as good now as when they were written; and as is observed by the writer in No. CXLI. of the Quarterly Review, in the amusing article on "Bee-books,"—"It would amply repay (and this is saying a great deal,) the most forgetful country gentleman to rub up his schoolboy Latin, for the sole pleasure he would derive from the perusal of the fourth Georgic." The aspect has been regarded as of the first importance; but there are points of greater consequence, namely the vicinity of good bee pasturage, the shelter of the hives from the winds by trees or houses, and their distance from ponds or rivers, as the high winds might dash the bees into the water.

Various aspects have been recommended, but the south, with a point to the east or west, according to its situation as respects the shelter it may receive from walls or trees, &c. is the best: care, however, must be taken that neither walls, trees, nor anything else impede the going forth of the bees to their pasturage.

"I have ever found it best," says Wildman, "to place the mouth of the hives to the west in spring, care being [pg 26] taken that they have the afternoon sun; the morning sun is extremely dangerous during the colder months, when its glare often tempts these industrious insects out to their ruin; whereas the mouth of the hive being then in the shade, the bees remain at home; and as clouds generally obscure the afternoon's sun at that season, the bees escape the temptation of going out. When food is to be obtained, the warmth of the air continues round the hive in the afternoon, which enables the bees to pursue their labours without danger.

A valley is a better situation for an apiary than a hill, being more convenient to the bees returning home with their loads; and, besides, bees are not so apt to fly away when swarming as when on a hill: but when swarms take a distant flight, they generally fly against the wind, so that the stragglers of the swarms may better hear the direction of the course taken by their fellow emigrants.

I recommend a hard gravel terrace for the hives to be placed upon, as being drier both in summer and winter for the bee-master to walk upon, when inspecting his bees, and also as less likely to afford shelter for ants or other enemies to bees; and, besides, it is better for the bees, which when much fatigued by their journeys, or benumbed by the cold, are apt to fall around the hives, and would recover more quickly from the warmth of the dry ground than if they had alighted on damp grass.

The hives should not be placed where water from the eaves of houses, from hedges, or trees, drop upon them; but they should be near the mansion house for the convenience of watching the bees, &c.

A small stream of water running near the hives is thought to be of advantage, especially in dry seasons, [pg 27] with gently declining banks, in order that the bees may have safe access to it.

Heaths, or places abounding in wild flowers, constitute the best neighbourhood for an apiary, and in default of this pasturage, there should be gardens where flowers are cultivated, and fields in which buck-wheat, clover, or sainfoin, is sown.

But cultivating small gardens of flowers for bees is useless, except a few early flowers near the hives for the bees to collect some pollen for the brood, such as the common kinds of crocus, white alyssum, single blue hepaticas, helleborus niger, and tussilago petasites, all of which flower early; but should any of the tribe of the willows grow near, there will be no necessity for cultivating the flowers abovementioned, as they yield an abundant harvest of farina, or pollen.

A rich corn country is well known to be a barren desert to the bees during a greater portion of the year. Hence the judicious practice of shifting the bees from place to place according to the circumstances of the season, and the custom of other nations in this particular well deserves our imitation.

Few places are so happily situated as to afford bees proper pasturage both in the beginning of the season and also the autumn; it was the advice of Celsus that, after the vernal pastures are consumed, they should be transported to places abounding with autumnal flowers; as was practised by conveying the bees from

Achaia to Attica, from Eubœa and the Cyclad Islands to Syrus, and also in Sicily, where they were brought to Hybla from other parts of the island.

Pliny states that the custom of removing bees from place to place for fresh pasturage was frequent in the [pg 28] Roman territories, and such is still the practice of the Italians who live near the banks of the Po, (the river which Pliny particularly instances,) mentioned by Alexander de Montfort, who says that the Italians treat their bees in nearly the same manner as the Egyptians did and still do; that they load boats with hives and convey them to the neighbourhood of the mountains of Piedmont; that in proportion as the bees gather in their harvest, the boats, by growing heavier, sink deeper into the water; and that the watermen determine from this, when their hives are loaded sufficiently, and it is time to carry them back to their places from which they came. The same author relates that the people of the country of Juliers used the same practice; for that, at a certain season of the year, they carried their bees to the foot of mountains that were covered with wild thyme.

M. Maillet, who was the French Consul in Egypt in 1692, says in his curious description of Egypt; "that in spite of the ignorance and rusticity which have got possession of that country, there yet remain in it several traces of the industry and skill of the ancient Egyptians." One of their most admirable contrivances is, the sending their bees annually into different districts to collect food, at a time when they could not find any at home.

About the end October, all such inhabitants of Lower Egypt, as have hives of bees, embark them on the Nile, and convey them up that river quite into Upper Egypt; observing to time it so that they arrive there just when the inundation is withdrawn, the lands have been sown, and the flowers begin to bud. The hives thus sent are marked and numbered by their respective owners, and placed pyramidically in boats prepared for the purpose. After they have remained some time at their furthest station, and are [pg 29] supposed to have gathered all the pollen and honey they could find in the fields within two or three leagues around, their conductors convey them in the same boats, two or three leagues lower down, and there leave the laborious insects so long a time as is necessary for them to collect all the riches of this spot. Thus the nearer they come to the place of their more permanent abode, they find the plants which afford them food, forward in proportion.

In fine, about the beginning of February, after having travelled through the whole length of Egypt, and gathered all the rich produce of the delightful banks of the Nile, they arrive at the mouth of that river, towards the ocean; from whence they had set out: care is taken to keep an exact register of every district from whence the hives were sent in the beginning of the season, of their numbers, of the names of the persons who sent them, and likewise of the mark or number of the boat in which they were placed.

Niebuhr saw upon the Nile, between Cairo and Damietta, a convoy of four thousand hives, in their transit from Upper Egypt to the Delta. Savary, in his letters on Egypt, also gives an account of the manner of transporting the hives down the Nile. In France floating bee-houses are common. Goldsmith describes from his own observation, a kind of floating apiary in some parts of France and Piedmont. "They have on board of one barge," he says, "three score or a hundred bee-hives, well defended from the inclemency of an accidental storm, and with these the owners float quietly down the stream: one bee-hive yields the proprietor a considerable income. Why," he adds, "a method similar to this has never been adopted in England where we have more gentle rivers, and more flowery banks, than in any part of the world, I know not; certainly it might be turned to advantage."

[pg 30]

They have also a method of transporting their hives by land in carts in Germany; and particularly in Hanover travelling caravans of bees may be seen during the season.

I have thus briefly quoted from famous authorities, to impress upon those who keep apiaries the importance of transporting their bees from pasture to pasture.

The advantage to weak swarms is very great, "but whilst so little of the true principles of bee management is understood, as that the destruction of the bees has been considered absolutely essential, in order to the attainment of their stores, it is no wonder that so little attention should have been paid to their cultivation in this country, and that it should not have proved a more productive department of rural economy."

"Bees, like everything else worth possessing, require care and attention; but persons generally think it is quite sufficient to procure a hive and a swarm, and set it down in the middle of a garden, and that streams of honey and money will forthwith flow; and, perhaps, commence calculating, from the perusal of the statements of the profits made by Thorley from a single hive, which he estimates to be 4300*l*. 16*s*. from 8192 hives kept during fourteen years! deducting ten shillings and sixpence, the cost of the first hive!"

The bar and frame-hives are so constructed that they can be moved from place to place with the greatest ease, and, perhaps, this may be an inducement for bee-masters to try the recommendations of transporting bees, and thus avoid one expense of feeding them during the winter.

Connected with the foregoing subject of transporting bees from place to place, is the question of the distance to which bees extend their flight in search of food, &c.; and the comparative excellence of the position of an apiary depends [pg 31] in some measure on the greater or less distance the bees will have to fly to their pasturage.

Dr. Chambers, and Dr. Hunter were of opinion, that the bee cannot extend its flight beyond a mile, which idea they adopted on the authority of Schirach; but then it must be recollected that the German mile of Schirach is equal to about 3-1/2 English miles.

It was the opinion of Huber, that the radii of the circle of the flight of the bee extended nearly to four English miles. And Huish says "The travelling apiaries of Germany, particularly those of Hanover, are regulated by the prevailing opinion, that the bee can, and does, extend its flight to four and even five miles; and acting upon that supposition, when the bee-masters move their apiaries, they always travel about two *stunden*, that is, about eight miles, as they then calculate that the bees are beyond the former range of their pasture by four miles." And adds, "a travelling apiary of 80 or 100 hives will exhaust the food within the area of a circle of four miles in about a fortnight or three weeks."

"But certainly there is no reason to fear that any part of this country will be overstocked with bees, for where one hive is now kept, fifty might be kept without running any risk of overstocking the country; for the average number of hives in the various apiaries does not exceed five."

"It has been calculated" says another authority, "that the pastures of Scotland could maintain as many bees as would produce 4,000,000 pints of honey, and 1,000,000 lbs. of wax; and were these quantities tripled for England and Ireland, the produce of the British empire would be 12,000,000 pints of honey, and 3,000,000 lbs. of wax per annum, worth about five shillings per pint for the honey, [pg 32] and one shilling and sixpence per lb. for the wax, making an annual produce in money of about 3,225,000*l*.

But in consequence of the present neglect of this branch of rural economy, we pay annually nearly 12,000*l*. for honey alone.

The imports and exports of wax bleached and unbleached were as follows:

	Imported.		Exported.		Returned for home Consumption.		the rate of Duty
	1831.	1832.	1831.	1832.	1831.	1832.	£ s. d.
Unbleached	7,005		1,878		10,002		1 10 0
Bleached	195 Cwt. 4,349		504 Cwt. 2,536		94	Cwt. 826	3 0 0

Produce of Duty.

| Unbleached | £ 10,262 |
| Bleached | 823 |

The price of wax varies (duty included) from 5*l*. to 10*l*. a cwt.

In 1831, 7,203 cwt. of wax were imported, of which 3,892 cwt. of it came from Western Africa; 1,551, from Tripoli, Barbary, &c.; and 910 cwt. from the United States.

In 1839, imports were 6,314 cwt., in 1841, 4,483 cwt. of wax; in 1838, 675 cwt. of honey; and in 1841, 3,761 cwt. valued at 12,000*l*. brought principally from the West Indies, Germany, and Portugal.

The above statement proves the demand there is in this country for honey and wax.

It is mentioned in Wildman's pamphlet that, when Corsica was subject to the Romans, a tribute was imposed upon it of no less than two hundred thousand pounds of wax yearly; but this is no proof of the excellence of their honey, which, according to Ovid, was of very ill account, and seems to be the reason why the tributary tax was exacted in wax, in preference to honey.

[pg 33]

The honey collected by the bees at all times retains qualities derived from the kind of plant from whence it has been procured, as is manifest not only by the peculiar odour of the honey, such as that collected from leek blossoms and all the onion tribe, but by the effects produced by the use of honey obtained from certain plants, chiefly from the subtribe Rhodoraceæ, such as the kalmia, azalea, rhododendron, &c., which yield a honey frequently poisonous and intoxicating, as has been proved by the fatal effects on persons in America. It is recorded by Xenophon in his Anabasis that, during the retreat of the ten thousand, the soldiers sucked some honeycombs in a place near Trebizonde, and in consequence became intoxicated, and did not recover their strength for three or four days; and these effects are supposed to have been produced from the honey having been extracted by the bees from the rhododendron ponticum or azalea pontica of Linnæus.

Although many of these plants have been introduced into this country, yet, probably from their small proportion to the whole of the flowers in bloom, the honey collected by the bees has not been found to be injured or to have produced any evil consequences.

The goodness and flavour of honey depend on the fragrance of the plants from which the bees collect it, and hence it is that the honey of different places is held in different degrees of estimation.

The honey gathered from the genus erica (termed *heather honey*) and most labiate plants, is wholesome. That which is made early in the year is preferred to what is collected in the latter end of the season. Whilst on the subject of honey, I will add the directions given by Wildman, how to separate the honey from the wax: "Take," he says, "the combs which have been extracted from the different [pg 34] hives or boxes into a close room, rather warm than otherwise, that the honey may drain more freely, and keep the doors and windows shut, to prevent the bees from entering, or else they will be very troublesome, and will attack and carry away the greater part of the honey from the combs.

"Lay aside such combs as have young bees or brood in them, as they would give your honey a bad flavour and render it unwholesome, and the bee-brood must also be separated and melted with the brood-combs. When you have thus separated the combs, let such as are very fine be nicely drained by themselves, without the least pressing whatever, having been carefully cleaned of every sort of filth, or insects, and dividing each comb in such a manner that the cells may be open at both ends, and placing them upon a sieve or coarse cloth, that the honey may drain off quite pure and undefiled. The remainder of the combs from which the honey has been thus drained, together with those which contained the bee-bread and brood, must be put into a coarse cloth or bag, and squeezed or pressed to get all the honey out. This will make it inferior in quality, and unfit for many uses, therefore it should be put into pots or bottles by itself, to feed bees with, for which purpose it will be better than pure honey, on account of the bee-bread that will be mixed with it, which is necessary for their subsistence.

"In order to obtain the wax in a pure state, what remains of the combs after separating the honey, together with the empty combs which had been laid aside, should be put into a copper with clean water; made to boil gently over a slow fire, keeping it constantly stirring. When it is melted, run it through a coarse cloth or bag made for the purpose, and put it into a press to separate the wax from the dross. Let the wax run from the press into a vessel placed under it, [pg 35] into which put some water to prevent the wax adhering to the sides.

"If this process of boiling and pressing is repeated twice or even three times, the wax will be much purer and consequently of greater value. Set it in a place where it may cool by degrees, in pans of the size you would choose your cakes to be, with some water in them, to prevent the wax sticking to the sides whilst hot. Honey should be kept only in stone jars, called Bristol ware, and in a cool and dry situation, but not corked up until a week or two after it has transuded through the sieve, &c., but should be carefully covered with perforated sheets of zinc to keep out insects and flies, &c. after which period the jars may be secured and put into the store-rooms.

"The only protection necessary for gentlemen,—for ladies, I presume, would never venture to undertake the dangerous task of extracting the honey combs from hives or boxes,—will be a pair of buckskin gloves, with a pair of worsted gloves over them extending to the elbows; so that the bees should not be able to creep between the gloves and the sleeves; for the face a piece of wire pattern gauze net, made in the shape of a bag, to draw with a string round the hat above the brim, which will keep it from the face, and the other open end being secured under the neck handkerchief, and with the assistance of a puff or two of smoke into any hive intended to be operated upon, the bee-master may fearlessly turn up the hive, and cut out combs or dislodge bees from their habitations, &c. with impunity."

[pg 36]

THE ENEMIES TO BEES, &c.

The proprietor having provided shelter for his bees, and as great a plenty of pasture as he possibly can, should next be careful to guard them from the numerous enemies which prey upon them, and destroy their honey-combs. Bees themselves, in the autumn and spring, are very often great enemies to one another, and rob each other's hives, especially in dry seasons, when the honey gathering is almost over; and the bees from over-stocked hives, not having honey sufficient for their winter's store, will through necessity attack the old hives or stocks, which are thinned by over swarming, carry away all their honey, and often destroy their queens. In order to prevent this havoc, contract the entrance or entrances of the hive attempted to be robbed, so that a few bees only can enter at a time, by which means the old stocks will be better able to defend themselves. If, notwithstanding this narrowness of the passage, robbers attack a hive, the entrance should be instantly closed and kept so till the thieves are gone, and it will be advisable in the evening to examine the state of the hive, especially as to weight, and if the queen be safe, remove it to another place, at least a mile from the old locality. The person who is thus employed, at a time when the bees are full of resentment, should be well defended from their stings. But, should he be so unfortunate as to get stung for his interference, the first thing is to extract the sting. To alleviate the irritation, cooling [pg 37] lotions should be applied, but the pain of a sting is relieved by applying spirits of hartshorn, or liquor potassæ, to the spot where the sting entered.

One would imagine the moth to be an enemy of no consequence, but the wax-moth (*Tinea mellonella*) is a most formidable enemy. She lays her eggs under the very skirts of the hive, or in the rubbish on the floor, or even in the combs of the bees; these eggs when hatched produce a small whitish worm or larva, and it is in this stage that it commits its ravages, extending its galleries through every quarter of the combs, detaching them from the tops and sides of the hives, and causing them to fall together.

The way to destroy them is frequently to lift up the hive in the morning, and kill all you can see. The most effectual way is to drive the bees into a new hive, but this can be only done in the height of the honey season; or the affected combs may be cut out, and the bees restored to their old habitation.

Mice are likewise very destructive to bees; sometimes they enter at the door, but most commonly near the top of the hive; this they do generally during winter, when the bees are in a torpid state; when this is suspected, set a few traps about the hives.

The common bat will also sometimes take possession of a hive, and commit very great havoc amongst the bees.

Wasps and hornets must be destroyed, if possible, either by gunpowder, or by the more primitive mode of placing limed twigs before the holes, when you have discovered their nests.

The spring is the time to kill the female wasps and hornets, for then, by the death of one female, a whole nest is destroyed. Or place bottles half full of sugar and beer where the wasps frequent; they will go in to drink, [pg 38] and drown themselves in the liquor, not being able to get out of the bottle again. Spiders must be killed, and their nets or webs broken down, otherwise they will catch and destroy many bees.

Swallows, frogs, ants, earwigs, snails, woodlice, poultry, and small birds of almost all kinds, are reckoned amongst their foes. And, therefore, there should be no lack of vigilance on the part of the owner of bees, to keep the bee-house as clean as possible from all vermin.

The signs of dysentery having commenced in any colony of bees may be known by the floor-boards and combs being covered with stains, by the dark coloured evacuations, producing an offensive smell, and frequent deaths amongst the bees. "Bees," says Gelieu, "have no real disease; they are always in good health as long as they are at liberty, are kept warm, and provided with plenty of food. All their pretended diseases are the result of cold, hunger, or the infection produced by a too close and long confinement during winter, and by exposure to damp, &c."

They appear however sometimes to be seized, in the spring, with dysentery; this is occasioned by their feeding too greedily, it is supposed, on honey dew, without the mixture of pollen and other wholesome nutriment.

The only remedy that has been found for this disease, is to give the bees plenty of honey, such honey as that extracted from the refuse combs in the autumn, that had abundance of bee-bread pressed amongst it,—the more the better,—mixing with it a tablespoonful of salt, and giving the bees their full liberty, and a clean hive. Many things are necessary for the preservation of bees, but more especially in this country, where the bees have only one season in five, on an average of years, really good for their honey harvest; wherefore the owner should take care [pg 39] to provide the light stocks with a sufficient quantity of food, which they have not been able to secure by their own industry, either through the badness of the bee-pasturage, the inclemency of the seasons, the weakness of the colony, or the spoil made by their enemies; and sometimes by the ill-judged management of their owners, in robbing the bees beyond the bounds of reason.

By this last unjust way of proceeding, these poor industrious little insects are absolutely starved, and their greedy masters deservedly experience the old proverb; that "Too much covetousness breaks the bag."

It is impossible to ascertain what quantity of honey will serve a hive of bees the whole winter, because the number in the hive may be more or less, and in some years, the spring is more forward than in others; but 25 lbs. is said to be the quantity required in a common cottage-hive. During frost, the bees consume very little food indeed; and still less during severe cold weather. Mr. White (with many other apiarians) is of opinion, that a greater degree of cold than is commonly imagined to be proper for bees is favourable to them in winter, for the bees during that period, are in so lethargic a state, that little food supports them.

The best method to feed the weak stocks, if in one of Mr. R. Golding's improved Grecian hives, is to place some combs (drone combs reserved for that purpose) filled on one side with honey, over the centre-board, and covering it over with a common hive.

The advantage of feeding bees from above is great; they are less likely to be attacked by the bees from other hives, and they do not become benumbed by the cold, as the same temperature is maintained above as in the rest of the hive.

But in all cases, bees should be fed in autumn, and [pg 40] before they are in absolute want of food, otherwise they will be so poor and weak that they will not be able to ascend or descend to feed themselves. When that happens, it is almost too late to save them; however, you may try and feed them, by first tying a piece of gauze over the bottom of the hive, turning it up to receive the heat of the sun or fire, and, if the bees revive at all, place a pewter dish with some liquid honey in it, on the floor-board, and the hive over it, when the bees will draw up the honey through the gauze or net without smearing themselves, the the pewter dish having been filled with hot water to keep the honey liquid, and to diffuse a genial warmth throughout the hive, and thus secure them for a time from the cold, which would chill and even kill the bees in the winter, when they came down to the bottom of the hive to feed on the proffered bounty.

In prosperous hives or colonies, as soon as the severity of the winter's frost is past, the queen-bee begins to lay her eggs in the various cells in the combs, and proceeds in proportion to the mildness of the season to deposit a succession. The number of young bees that may by this means rise in a hive, may endanger the lives of all the bees by famine, for the increased multitude consume a great deal of honey, an accident likely to happen if the mild weather of January or February should be succeeded by cold, rainy, or even dry weather; for it is found that the flowers do not secrete the sweet juices, which constitute honey, so freely during the prevalence of dry easterly winds; and thus present a barren field for the out-of-door labours of the bees.

On this account, the proprietor should examine the hives frequently at this season, that, if necessary, he may give them a proper supply, in which he should be bountiful rather [pg 41] than otherwise, because the bees are faithful stewards, and will return with interest what is thus in their great need bestowed upon them.

The time of the bees' swarming is generally in the months of May and June, and sometimes July, but the latter is too late, as there are then fewer bees than in the earlier swarms, and they seldom live through the winter without much care and feeding.

The later swarms should be hived in rather smaller hives than the first, that, by clustering together, they may the better nourish and keep themselves warm.

The hours of their swarming are for the most part about twelve o'clock at noon, never before eight, and seldom after four in the afternoon.

The symptom of swarming, is generally the unusual number of bees seen hanging at the mouth of the hive, and if a piping noise, or a shrill note, which is made by the queen is heard, it is a sure index the bees will swarm, if the weather be warm and dry.

If the bees work a comb under the floor-board, as is sometimes the case, it is a sign they will not swarm; a more certain sign is when they throw out the young dead queens with the drone brood. When they retain the drones in the hives after August, it is a bad omen, as they are then reserved for the sake of the young queens, which they are expecting to raise; and the season being too far advanced, and their failing in the attempt, and being without a queen, the colony will most certainly dwindle away, before the next season.

Always choose a hive proportionable to the size of your swarm, and prepare to hive them as soon as possible, lest they should rise again. It is not unusual to ring a bell or tinkle a brass pan, &c., at the time the bees swarm; it is [pg 42] also a common method to dress the hives with honey, balm, &c.

I mention these things, because they are customs of long standing: the tinkling of bells is of little use, as the bees will generally settle near the hive; and as to dressing the hives, I by no means recommend it, as the bees like a clean new hive much better, for it does not give them so much trouble to clean, &c.

If the swarm should rise in the full heat of the day, and the sun shine hot upon them, they will not continue long in their first situation; for when they find they have all got their company together, they will soon uncluster, rise again, fly to some particular spot which has been fixed upon for that purpose by detached parties of bees, who return and acquaint the swarm; therefore I would advise to hive them as soon as possible, and remove them in the evening to the place where they are to remain.

The supposed relative value of early and late swarms is thus mentioned in an old English proverb:—

>A swarm in May,
>
>Is worth a load of hay.
>
>A swarm in June,
>
>Is worth a silver spoon;
>
>A swarm in July,
>
>Is not worth a fly.

[pg 43]

SWARMING AND HIVING THE BEES.

Every good swarm should weigh about 5 lbs., and according to the account given in Key's Treatise, would contain 23,000 bees. The manner of hiving them must be regulated chiefly by the places upon which they alight.

If they settle on a dead hedge, or upon the ground, set a hive over them, putting props under it if necessary, and, with a large spoon or brush of wet weeds, stir them softly underneath, and they will go in.

If they should happen to settle upon a small bough, you may cut it off, and laying it quietly on a cloth, place a hive over them; or if you cannot conveniently separate the bough from the body of the tree, you may shake or sweep them off into the hive.

If the sun shines hot upon it, shade it with a few boughs, &c., but let it remain near the place where the bees settled until the evening, at which time move it to the bee-house, or the place where it is to stand during the season, as just directed.

If the bees have hung a considerable time to the place where they first settled, you will, perhaps, find it difficult entirely to dislodge them, as they will neglect their labour and fly about the spot for many days afterwards. The best method to prevent this is, by rubbing the branches with rue, or any kind of herb disagreeable to the bees; but be careful not to hurt any of the bees.

Swarms seldom return home again, when they are well [pg 44] settled, and if you find them inclined to do so, depend upon it, some accident has happened to their queen, which you will easily ascertain by their making a murmuring noise, and running in a distracted manner over and about the sides of the hive. When you observe this, immediately seek about for her, beginning with the stock-hive from whence the swarm rose, and pursue the track they took at setting out; you will seldom miss finding her, for she is never alone, but generally encompassed with a cluster of bees, who would sooner perish than leave her in danger.

When you have found her, take her up gently, and put her to the swarm, and you will soon find the cause of their dissatisfaction removed by the arrival of the queen.

The greatest care must be taken to have your hive clean and sweet, free from loose straws or other obstacles, which will create great trouble and loss of time to the bees, if left to them to remove.

If bees have flowers suitable to their tastes, and no great distance to travel to them, they will fill their hives both with honey and wax, in about a month or five weeks, and, if the season has proved fair and pleasant, in less time; but the bee-keeper must expect four out of every five seasons to be unpropitious to his little charge, and, therefore, he must be on the watch to assist them with food in the time of need.

Scarcely has the swarm arrived at its new habitation, when the working bees labour with the utmost diligence, to procure food and build their combs. Their principal aim is not only to have cells in which they may deposit the honey and pollen, but a stronger motive seems to animate them; they seem to know that their queen is about to deposit her eggs; and their industry is such, that in four and twenty hours they will have made combs, twelve inches [pg 45] long, and three or four inches wide. They build more combs during the first fortnight, than they do during all the rest of the year.

Other bees are at the same time busy in stopping all the holes and crevices they happen to find in their new hive, in order to guard against the entrance of insects which covet their honey, their wax, or themselves; and also to exclude the cold air; for it is indispensably necessary that they be lodged warm and secure from damp, &c.

A second swarm scarcely is, and much less are the third ones called *casts* worth keeping single, because, being few in number, they cannot allow so large a proportion of working bees to go abroad in search of provisions, as more numerous swarms can, after retaining a proper number for the various works to be done within the hive.

Bees sometimes swarm so often that the mother-hive is too much weakened or reduced in population. In this case they should be restored; and this should also be done when a swarm produces a swarm the first summer, as is sometimes the case in early seasons.

The best way, indeed, is to prevent such swarming, by giving the bees more room; though this, again, will not answer where there is a prolific young queen in the hive; as she well knows that her life is the forfeit of her remaining at home.

Before the union of one or two casts or late swarms is made, it is better to kill one of the queens, if possible, to prevent the queens destroying one another.

If an old hive is full of bees, and yet shows no disposition to swarm, puff in a little smoke at the entrance of the hive, then turn the hive up, and give it some slight strokes on the sides so as to alarm the bees. They will immediately run to the extremities of the combs, and if you [pg 46] then attentively examine them, you will, in all probability, perceive the queen-bee the foremost amongst them. Seize her between your fore finger and thumb, and confine her in your hand till most part of the bees take wing; let her then go, the bees will soon join her, and settle on some branch of a tree. Put them into an empty hive. Restore the old hive in its place, that the bees which have been out in the fields may enter it on their return, and having allowed them to remain there an hour or two, place it upon another stand near or next to its own.

The hive having what may now be called a swarm in it, is then placed on the stand of the old stock; and if the bees in both hives work regularly, carrying in loads of pollen on their thighs, all is well.

Bees are not apt to sting when they swarm naturally, therefore, it is not necessary then to take extraordinary precaution against them; but when any of these violent and artificial modes are attempted, I should advise the operator to be well guarded at all points.

Wildman weighed bees and found it required 4,928 bees to make a pound of sixteen ounces, but the different circumstances in bees may occasion a considerable difference in their weight. When the bees swarm, they come out loaded with wax secreted in their wax pockets and honey in their honey bags, and would weigh heavier than bees taken for that purpose by chance; and, therefore, the number of the bees is not to be thus computed, from the weight of the swarm; for one fourth of the number at least should be deducted, in lieu of the wax and honey they have brought off with them. There is also another allowance to be made, namely, that when alive, they do not probably weigh so heavy as when dead.

The person who intends to erect an apiary, should purchase [pg 47] a proper number of hives at the latter end of the year, when they are cheapest. The hives should be full of combs, and well stored with bees.

The purchaser should examine the combs, in order to know the age of the hives. The combs of that season are white, those of a darkish yellow are of the previous year; and, where the combs are black, the hives should be rejected, because old hives are most liable to vermin and other accidents.

If the number of hives wanted were not purchased in the autumn, it will be necessary to remedy this neglect after the severity of the cold is past in the spring. At this season, bees which are in good condition, will get into the fields early in the morning, return loaded, enter boldly, and do not come out of the hive in bad weather; for when they do, this indicates that they are in great want of provisions.

They are alert on the least disturbance; and by the loudness of their humming, you can judge of their strength. They preserve their hives free from filth, and are ready to defend it against every enemy that approaches.

But the better plan is at once to commence with new hives, and purchase the first and strong swarms to put into them, and introduce them into the bee-house.

There are various substances found in a hive, such as the *wax*, with which the combs are built, the *honey*, the *farina* or *pollen*, with which the bee-brood is fed, and *propolis*.

Honey, is a fluid or semi-fluid substance, the materials of which are collected by the bees, from the nectaries at the base of the corollæ of flowers, where this vegetable production is secreted.

It cannot be said to be a purely vegetable production when found in the combs, for after being collected by the [pg 48] insect by means of its proboscis, it is transmitted into what is called the honey bag, where it is elaborated, and, hurrying homewards with its precious load, the bee regurgitates it into the cell of the honey comb. It takes a great many drops to fill a cell, as the honey bag when full does not exceed the size of a small pea.

When the cell is full, it is sealed up with a mixture of of wax and pollen, and reserved for future use in winter and spring.

Wax. There are several varieties of this substance, but bees-wax is a secretion of that insect from its ventral scales. With this substance the comb is constructed; it takes the bees, according to Huber's account, twenty-four hours to secrete the six laminæ of wax in the wax pockets, which may be seen to exude between the segments of the under side of the abdomen of the bee. For the purpose of the formation of wax, the bees have to cluster and form themselves into festoons from the top of the hive, and after the elapse of the necessary period, the wax scales are formed, with which the bees commence immediately to build their combs, and the various cells for the reception of the brood or food, according to the season of the year.

Propolis, is a tenacious, semi-transparent substance, having a balsamic odour; which the bees gather from the buds of certain trees in the spring, such as the horse-chestnut, the willow, the poplar, and the birch.

This tenacious substance is employed by the bees to attach more firmly the combs to the top or foundation, and also the edges of the combs to the sides of the hive or box, to stop the crevices, and fasten the hives or boxes to the floor-boards, and in forming barriers against the intrusion of enemies.

[pg 49]

Farina, or *Pollen*, is the dust or minute globules contained in the anthers of flowers, and is the fertilizing property of flowers, which the bees thus assist to carry, whilst travelling from flower to flower, without which the flowers would not fructify. The bees have been found to continue collecting pollen from the same species of flowers, and prevent the multiplication of hybrid plants. They collect and carry this substance on the outer surface of the tibia, or the middle joint of the hinder leg; this part of the leg is broad, and on one side it is concave, and furnished with a row of strong hairs on its margins, forming as it were a natural basket, well adapted for the purpose. This substance mixed with honey, forms the food of the larvæ or young brood, after undergoing, perhaps, a peculiar elaboration by the working or nurse bees.

Having thus mentioned the different substances found in a hive, it only remains to add a short history of the inmates of the hive. Every swarm of bees comprises three distinct kinds of the same species, namely, the *female* or *queen*, the *neuter* or *worker-bee*, and the *male* or *drone*.

As there is only one *queen-bee* in each swarm or colony, she is seldom to be seen amidst the thousands of other bees; but she is easily distinguished from the rest by her slower movements, her greater length and larger size; and the general appearance of her body, being of a more dark orange colour, and her hinder legs having neither brushes nor pollen baskets upon them, although longer than those of the worker-bee; her wings also appear stronger, and she possesses a more curved sting, which she seldom uses, except when asserting her rights to the sovereignty of the hive.

Without a *queen-bee* no swarm can thrive, for she is not only the ruler, but chiefly the mother of the community in [pg 50] which she dwells, and wherever she goes, the greatest attention is paid her. In the hive, the utmost solicitude is evinced to satisfy her in every wish; wherever she moves the bees anxiously clear away before her, and turn their heads towards their sovereign, and with much affection touch her with their antennæ, and supply her, as often as she needs, with honey or other delicacy which their own exertions, or those of their fellow labourers, have gathered for her use.

The queen-bee is said to live four or five years, and is generally succeeded on her throne by one of her own descendants duly brought up for the purpose; but in the event of her untimely decease, the workers have the power of raising a sovereign from amongst themselves, and fitting her for the station she is intended to occupy; this they do by selecting one of the larvæ of the worker-bee of a certain age, and, enlarging the cell which it is to occupy, supplying it with a nourishment different from that which they give to the worker and drone-brood.

A *queen-bee* takes seventeen days to arrive at maturity, that is to say, from the egg-state to the fully developed queen, but this period will vary as a sudden change of temperature will prolong the interval; and this also applies to the perfect *queen* herself, who will not deposit her eggs in the cells, when any severe weather happens at the period she may be expected to produce the eggs.

The fecundity of the queen-bee is very great, for it is estimated that during breeding time, unless prevented by the cold weather, she lays at the rate of from one hundred to two hundred eggs a day, causing an increase of not less than eighty thousand worker-bees, and drones included, in a season when circumstances are favourable.

The cells formed for the royal brood are very different [pg 51] from those of the males or the workers, and are generally suspended from the sides or edges of the combs; in shape they are very much like a pear, the thickest end joining the comb, and the small end having the mouth or entrance to the cell, and hanging downwards, and being almost as large as a lady's thimble.

The *drones* or *males* in a hive are computed at from six hundred to two thousand, but the numbers are remarkably irregular, and the proportion is not regulated by the number of bees contained in a hive; for a small swarm or colony will contain as many, or more sometimes, than a large one.

The drone may be easily distinguished from the *queen* or *workers*, from its greater breadth, having large eyes which meet at the top of the head, and no sting, and from its making a loud humming whilst flying.

It takes twenty-four days from the time of the laying of the drone *egg* to its coming forth a perfect insect. Drones are generally hatched about the end of April or the beginning of May; they venture out of the hive only in warm weather, and then only in the middle of the day, and they are generally expelled by the bees from the hives about July or August, after the impregnation of the young queens has taken place.

When the destruction of the drones takes place earlier, it may be considered a certain indication that no swarming will take place during that season; but the retention of the drones after August, is a very bad sign, as the swarm must certainly perish in the winter, unless their vacant throne is supplied with a prolific queen.

The *neuter* or *worker-bee*, is the least of the three, and of a dark brown colour; the abdomen is conical, and composed of six distinct segments, and armed with a straight sting; [pg 52] it possesses a long flexible trunk, known by the name of a proboscis, and has on its two hinder legs a hollow or basket, to receive the propolis and farina which it collects as before described.

The number of workers in a well-stocked hive is about fifteen thousand or twenty thousand. Upon them devolves the whole care and labour of the colony, to collect pollen, propolis, and honey; to build the combs and to attend upon the brood or young bees.

The *worker-bee* is short-lived, seldom surviving more than a year, but this is more from the toil they have to endure, though it be a labour of love, and the many risks they run upon each occasion of going out in search of food, &c., from the weather, or their numerous winged enemies.

> "Sunt quibus ad portas cecidit custodia sorti:
>
> Inque vicem speculantur aquas et nubila cœli,
>
> Aut onera accipiunt venientum, aut agmine facto
>
> Ignavum fucos pecus à præsepibus arcent.
>
> Fervet opus, redolentque thymo fragrantia mella."

www.ingramcontent.com/pod-product-compliance
Lightning Source LLC
Chambersburg PA
CBHW030511220526
45464CB00006B/2745